JOSHUA CHAPMANN

Machine Learning

Fundamental Algorithms for Supervised and Unsupervised Learning With Real-World Applications

JOSHUA CHAPMANN

D1382302

i

Chapter 2 – Supervised vs Unsupervised Learning Algorithms

If you want a machine learning algorithm to solve a problem, you must first train it. Unlike humans, you do not teach a machine in a classroom; you teach machines using **training data**. Essentially, training data are past examples showing the machine how to solve a problem. Using your chosen algorithm, the machine works through all these examples and *understands* how to solve them. If you use a few very specific examples, the machine will only be able to solve very similar problems. If you use very large and broad training data, your machine will learn how to solve a wide range of diverse problems.

Think of yourself as the machine and training data as textbooks. If you read a lot of textbooks you will gain a lot of knowledge, which you can use to solve a very wide range of problems. You can make connection between the different books and attempt challenges you have never seen before. If you only read a few, short books you do not have much knowledge. You can only solve a few, very similar problems to what you just read. Machine Learning is exactly like that!

Although there are many vastly different machine learning algorithms you can use, these can all be grouped into two main classes: **supervised learning** and **unsupervised learning.**

Supervised Learning

The only difference between supervised and unsupervised learning lies in the type of training data you use. In supervised learning you provide **labelled** training data. This means that each example you feed into your algorithm is classified into a recognizable data class or type. If this is the first time reading about labelled training data, it may appear confusing. Do not panic, an example will make everything clear once again:

Think of yourself trying to learn a new language, let's say Italian; your teacher gives you training data in form of vocabulary. This data is labelled because each new word you learn is classified into recognizable terms. For instance, your teacher tells you "ciao" means "hello" in English, she tells you "mangiare" means "eating", etc... Each example of a new Italian word has been associated into an English word you can understand. These are the fundamentals of supervised learning and machines work the exact same way.

Unsupervised Learning

As humans, unsupervised learning is a trickier concept to understand because we do it subconsciously. You can think of it as learning through observation.

Unsupervised learning is defined by **unclassified** training data. Essentially, you provide a machine with thousands of problems and their results, but you do not explain how the result is calculated. The algorithm then starts to look for things in common; it tries to identify shared traits and features between the problems and the solutions. With enough data, the algorithm can extracts a pattern and develop a strategy to solve any problem. Again, these concepts are best explained using practical examples:

For instance, let's jump back to how you can learn a new language – Italian. An example of how you can learn Italian using unsupervised learning is by watching movies. If you don't know any Italian and you watch an Italian movie you will not understand anything. Second movie – nothing, third movie – still nothing. After ten movies you might start to recognise common expressions, for instance every time a character says "Arrivederci" you observe someone leave and you conclude this means "goodbye".

If you watched thousands of movies, you would learn how to speak Italian perfectly using this technique, you would simply "pick it up".

Clearly, you can't do that because it's impossible; it would take up too much time for you- **but not for a computer.** You could acquire millions of hours of Italian speech and run them all through a machine learning algorithm. After "listening and learning" to all these hours of Italian, the computer will have the meaning of the language perfectly.

Chapter 3 – Supervised Learning: K-Nearest Neighbor

Following four years of my career in analytics, more than 80% of the models I have built have been classification models whereas 15-20% have been regression models. Throughout the industry, you are able to generalize these ratios. A reason for the heavy bias towards classification models is that many analytical problems involve making a clear decision. As an example, we could ask whether a custom will attrite or not, whether we should target customer X for digital campaigns and whether a particular custom has high potential or not. As you can see, you can create many insightful analysis and directly link them to an implementation roadmap.

An interesting classification technique you can use is known as the K-nearest neighbours (KNN) We will be primarily focusing on how this algorithm works as well as how it does the input parameter affect the output/prediction.

When to Use KNN Algorithm?
There are three important aspects that you will need to consider in order to determine whether to use KNN. The KNN algorithm can be used for both classification and regression predictive problems although it is more widely used in industry classification problems. The three aspects that you will need to consider are:

1. Ease of Interpreting Output

2. Time of Calculation

3. Predictive Power

Here are a few examples that we can place KNN in the scale:

	Logistic Regression	CART	Random Forest	KNN
1. Ease to interpret output	2	3	1	3
2. Calculation time	3	2	1	3
3. Predictive Power	2	2	3	2

As you can see the KNN algorithm fairs across all the parameters that we considered. From this you can also see how the KNN algorithm is commonly used due to the fact that is easy to interpret and the calculation time is quite low.

How does the KNN Algorithm Work?

We will use a case study that is quite simple to understand in order to demonstrate how the KNN Algorithm works. The following is a spread of red circles (RC) and green squares (GS):

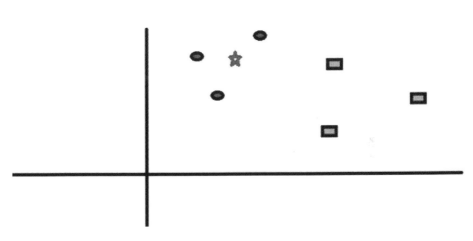

Now we want to find out the class of the blue square (BS). We know from looking at the graph that the BS can either be RC or GS and nothing else. The "K" in KNN algorithm we are able to take vote from. For example, let's say that K was to equal 3. From there we can make the circle with BS as centre and only as big to encapsulate the three data points on the plane as you can see from the diagram:

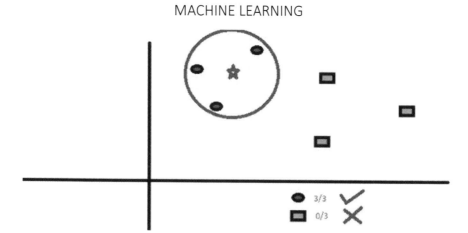

As you can see from the three closest points to BS, they are all RC. Therefore, we can be confident that the BS should fall into the class of RC. The decision becomes quite clear once all three of the votes had gone to RC. The determining choice of K is quite fundamental in this algorithm. Our next step is to determine how we are able to deduce what is the best K and the factors that go into this decision.

How do We Choose the Factor K?

In order to determine how we would choose the factor K, we will first need to have an understanding of exactly how K influences our algorithm. From our last example, we could see that all 6 of our training observations remained constant. The given value of K allows us to make boundaries of each class. The boundaries that are made also allow us to segregate RC from GS. In similar way, we can see the effect of the value K on the class boundaries. Here are a number of different boundaries that show the separating of the two classes with different values of K.

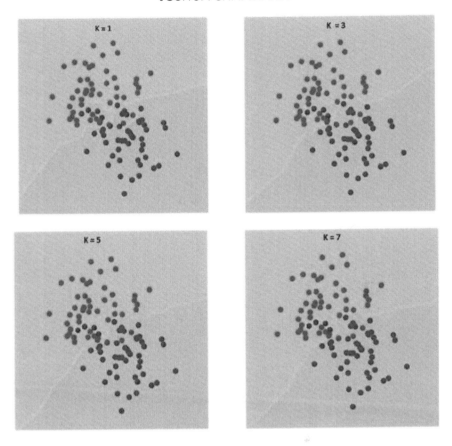

While observing these images, you are able to notice the boundary becoming smoother the value of K increasing. The value of K begins to increase into infinity and as it does it becomes all blue or all red, depending on the total majority. We will need to access two parameters on different K value, these being the training error rate and the validation error rate. You are able to see this through the following cure, the training error rate with varying value of K:

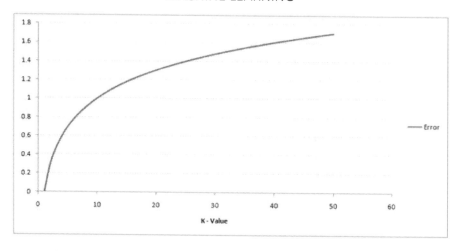

From this graph, we can see that the error rate at K=1 is a constant zero for the training sample. The reason for this is that the closest point to any training data is in fact the point itself giving a value of K=1. In the case of the validation error curve, we would have a similar result with K also having a value of 1. To demonstrate this, below is a validation error curve with a varying value of K.

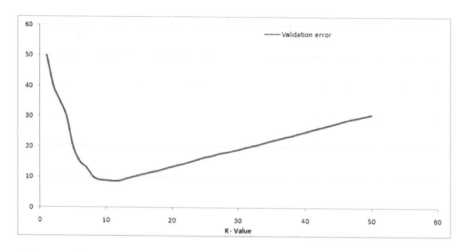

We are able to have a much clear image from observing this graph. We were overfitting the boundaries at K=1 therefore you will notice that the error rate will initially decrease and reach a minimum.

Following these minima point, it will then increase as the value of K increases. In order to reach the optimal value of K, we will need to separate both the training and validation from our original dataset. From there you will need to plot the validation error curve in order to get the optimal value for K. Once you have reached this value, you are able to use this value of k for all future predictions.

End Notes

The KNN algorithm is one of the simpler classification algorithms you can learn. However, despite this simplicity, you may still attain highly competitive results. You are also able to use the KNN algorithm for regression problems. There is a difference however. The methodology we discussed is used to determine averages of the nearest neighbours rather than voting from nearest neighbours. You are able to use KNN to code as a single line on R.

Chapter 4 – Supervised Learning: Naive Bayes

Imagine for a moment that you are working through a classification problem and from that you have generated your own set of hypotheses and designed features as well as a discussion around the importance of the variables. Your stakeholders reach out to you and request that you show a first cut of the model you are producing. With hundreds of thousands of data points to show as well as numerous variables throughout your training set, you are at a loss for what to do. There is a one way in which you are able to demonstrate your model at this early stage and that is with the 'Naive Bayes' algorithm. In this chapter, we will go through the basics of the Naive Bayes algorithm as it is quite useful in situations where you are faced with large data sets.

What is Naive Bayes Algorithm

The Naive Bayes algorithm is a technique based on the Bayes' Theorem which is used for classification. The technique utilizes an assumption of independence among predictors. To put things simply, the Naive Bayes classifier assumes that the presence of a feature class is unrelated to the other features in the class. Using an apple as an example, we can say that it has features of being red, round and roughly 3 inches in diameter. While these features may depend on each other or the existence of other features, they each have properties that would indicate that the fruit we have is an apple, this is how we come up with the name "naive".

The model is really quite simple to put together and can significantly help when it comes to very large data sets. The advantage of the Naive Bayes model is that it is not only extremely simple, it can even outperform some of the more sophisticated classifications methods.

The Bayes theorem allows for a way to calculate posterior probability $P(c|x)$ from $P(c)$, $P(x)$ and $P(x|c)$. Referring to the equation below:

$$P(c|x) = \frac{P(x|c)P(c)}{P(x)}$$

Likelihood — Class Prior Probability — Posterior Probability — Predictor Prior Probability

$$P(c|X) = P(x_1|c) \times P(x_2|c) \times \cdots \times P(x_n|c) \times P(c)$$

- $P(c|x)$ is the posterior probability of class (c, target) given predictor (x, attributes).
- $P(c)$ is the prior probability of class.
- $P(x|c)$ is the likelihood which is the probability of predictor given class.
- $P(x)$ is the prior probability of predictor.

How Naive Bayes Algorithm Works

Using the algorithm in an example, refer to the data set below where we have a set of weather and corresponding target variable, in this case whether or not it is possible we will 'Play'. From here we will match up the data with whether or not the weather conditions will permit players to play. In order to do this, we will need to take the following steps:

Step 1: We will need to take this data and convert it over to a frequency table

Step 2: We will then need to create a Likelihood table by determine the probability such as Overcase = 0.29 and the probability of playing is 0.64.

Weather	Play
Sunny	No
Overcast	Yes
Rainy	Yes
Sunny	Yes
Sunny	Yes
Overcast	Yes
Rainy	No
Rainy	No
Sunny	Yes
Rainy	Yes
Sunny	No
Overcast	Yes
Overcast	Yes
Rainy	No

Frequency Table		
Weather	No	Yes
Overcast		4
Rainy	3	2
Sunny	2	3
Grand Total	5	9

Likelihood table				
Weather	No	Yes		
Overcast		4	=4/14	0.29
Rainy	3	2	=5/14	0.36
Sunny	2	3	=5/14	0.36
All	5	9		
	=5/14	=9/14		
	0.36	0.64		

Step 3: Now it is time to use our Native Bayes equation in order to calculate the posterior probability for each our classes. The class revealed to have the high posterior probability is on the one determined to be the outcome of our prediction.

Problem: Is it correct to state that Players will play if weather is sunny?

We are able to solve this using the method we had discussed to determine the posterior probability.

P(Yes | Sunny) = P(Sunny | Yes) * P(Yes) / P (Sunny)
Here we have P (Sunny |Yes) = 3/9 = 0.33, P(Sunny) = 5/14 = 0.36, P(Yes)= 9/14 = 0.64
Now, P (Yes | Sunny) = 0.33 * 0.64 / 0.36 = 0.60, which has higher probability.

In order to predict the probability of different classes based on varying attributes, Naive Bayes has a similar method which is employed. You will notice that the algorithm is mainly used in text classification as well as problems which involve multiple classes.

What are the Pros and Cons of Naive Bayes?

Pros:

- Simple and quick to predict class of test data set. The technique performs well in cases where their multiple classes

- The Naive Bayes classifier performs much better compared to other models when the assumption of independence holds. Involves less training data than models such as logistic regression.

- When categorical input variables are involved, the model performs quite well in contrast to numerical variables. When numerical variables are involved, normal distribution is assumed.

Cons:

- In cases where the categorical variable has a category in the test data said which has not been observed in the training data set, the model will assign a 0 probability in which case is unable to make a prediction. This is identified as the zero frequency. There is a way to solve this and this is called a smoothing technique. One smoothing technique which you can use is called the Laplace estimation

- Naive Bayes has been known as a bad estimation meaning the probability outputs should not be taken too seriously.

- Naive Bayes is limited in the fact that the assumed independent predictors are almost impossible to get.

Applications of Naive Bayes Algorithms

- Real Time Prediction: As naive bayes is fast and eager learning classifier, it has huge potential for making predictions in real time.

- Multi Class Prediction: Since the algorithm is particularly adept at making multi class predictions, the Naive Bayes algorithm is able to predict the probability of multiple classes of a target variable.

- Text Classification/ Spam Filtering/ Sentiment Analysis: As Naive Bayes classifiers are mostly used in text classification, they have a higher success rate in comparison with other algorithms. The reason they are mainly used in text classification is too due to the fact that they produce better results in multi class problems and independence rule. What this means is that they are able to be used in Spam filtering by identifying spam emails as well as sentiment analysis such as in social media in order to identify both positive and negative customer sentiments.

- Recommendation System: Combining both the Naive Bayes Classifier and Collaborative filtering, you are able to build a recommendation system that makes use of machine learning as well as data mining techniques allowing unseen information to be filtered as it predicts whether you would prefer a particular a resource or not.

Building a Basic Model Using Naive Bayes in Python

Using python library, you can build a Naive Bayes model in Python. There are three types of Naive Baye model you are able to build under scikit learn library, these are Gaussian, Multinomial and Bernoulli. Gaussian is used in classification, assuming that the features will be following a normal distribution. Multinomial is more used for discrete counts. As an example, if we were to be working with a text classification problem, we can consider bernoulli trials, taking one step further and rather than working with the word occurring in the document we would like to count how often the word occurs in the document. You can think of this as the number of times outcome number x is observed over the n trials. Bernoulli is used when your feature vectors are binary, comprised of zeros and ones. A practical application of Bernoulli is for when the application would be text classification with 'bag of words' model where the 1s and 0s represent where the word occurs in the documents and or whether it does not occur in the document respectively.

Chapter 5 – Supervised Learning: Logistic Regression

Introduction

Machine learning thrives under a specific set of conditions. In this case you will need to make sure that your algorithm fits with the assumptions and requires that ensures superior performance. You won't be able to use any algorithm in just about any condition. As an example, you could not use linear regression on a categorical dependent variable. This simply won't work as the values will be extremely low for adjusted R2 and F statistic.

In these kinds of situation s, it is best to try using algorithms such as Logistic Regression, Decisions Trees, SVM, Random Forest and others. In order to discover more on these algorithms and have a basic overview, you can explore "Essentials of Machine Learning Algorithms". Following this book, this is a great place to continue learning as this chapter will be exploring Logistic Regression in R. Once you have mastered linear regression, you should continue to learn some of the more advanced algorithms. In order to have a full understanding, we will be covering the science behind this algorithm.

What is Logistic Regression?

As a classification algorithm, Logistic Regression is used to predict a binary outcome such as 1 or 0, Yes or No, True or False depending on a set of given independent variable. In order to represent our binary outcome, we will be using dummy variables. Logical regression can also be thought of as a special case of linear regression when the outcome variable is categorical and where we are using log of odds as dependent variable. To put things simply, we are able to predict the probability of an event occurring by fitting that data into a logit function.

Derivation of Logistic Regression Equation

Logistic Regression forms part of a much larger class of algorithms that are known as Generalised Linear Model (GLM). Nelder and Wedderburn proposed the model back in 1972 in order to provide a means of using linear regression to the problems which linear regression was not suited towards. Nelder and Wedderburn were able to propose a class of different models, these being linear regression, ANOVA, Poisson Regression and others, logistic regression was included as a special case. The fundamental equation of generalized linear model is: $g(E(y)) = \alpha + \beta x1 + \gamma x2$

In this case g () is the link function, E(y) is the expectation of target variable and $\alpha + \beta x1 + \gamma x2$ is the linear predictor. We will use the link function in this case to link the expectation of y to linear predictor.

There are some important points to consider here. GLM does not assume a linear relationship between dependant and independent variables. What is does assume however is that a linear relationship between link function and independent variables exists in the logit model. Furthermore, the dependent variable does not need to be normally distributed. The equation does not use Ordinary Least Square or OLS for parameter estimation. Rather it utilises maximum likelihood estimation (MLE). Finally, any errors need to be independent but not normally distributed.

We can use an example to better illustrate how the equation works. If we were to have a sample of 1,000 customers and we needed to predict the probability of whether a custom would buy (y) a particular magazine or not, we could use logistic regression in order to reach an outcome. The reason for this is that we have a categorical outcome variable. In order to start the logistic regression, we will need to take down the simple linear regression equation using a dependent variable enclosed in a link function.

$g(y) = \beta o + \beta(Age)$ ---- (a)

In order to make this easier to understand, age has been considered an independent variable.

Since we are using logistic regression, we are only concerned about the probability of outcome dependent variable, whether success or failure. As mentioned previously, g() is the link function. We are able to reach this conclusion using two things: The probability of success (p) and the probability of failure (1-p). P should then meet the following criteria:

1. Must have a positive value as p >=0

2. Must be less than equal to 1 as p >=1

In order to get to the core of logistic regression we will need to satisfy these 2 simple conditions. By denoting g() with 'p' to begin with we will be able to establish the link function. Now probability will always be positive so we will need to put the linear equation in exponential form. With any value of slope and dependent variable, we will never receive a negative exponent from this equation.

$$p = \exp(\beta o + \beta(Age)) = e^{\wedge}(\beta o + \beta(Age)) \quad \text{------- (b)}$$

In order to come to a probability less than 1, we will divide p by a number greater than p. We can do this by:

$$p = \exp(\beta o + \beta(Age)) / \exp(\beta o + \beta(Age)) + 1 = e^{\wedge}(\beta o + \beta(Age)) / e^{\wedge}(\beta o + \beta(Age)) + 1 \quad \text{----- (c)}$$

From here we can use (a), (b) and (c) in order to redefine the probability as:

$p = e^{\wedge}y/\ 1 + e^{\wedge}y$ --- (d)

Here p is the probability of success. This (d) is the Logit Function. Therefore, if p is the probability of success, 1-p will be the probability of failure, we can write it as this:

$q = 1 - p = 1 - (e^{\wedge}y/\ 1 + e^{\wedge}y)$ --- (e)

In this case q is the probability of failure.

When we divide (d) / (e), we get:

$$\frac{p}{1-p} = e^y$$

We then take log on both sides to get:

$$\log\left(\frac{p}{1-p}\right) = y$$

Log(p/1-p) is therefore the link function. Logarithmic transformation on our outcome variable enables us to model a nonlinear association in a linear way.

By substituting the value of y, we receive:

$$\log\left(\frac{p}{1-p}\right) = \beta_0 + \beta(Age)$$

We are able to use this equation in Logistic Regression. In this case (p/1-p) is the odd ratio. In the case that an odd ratio is found to be positive, there is a probability of success greater than 50%. We demonstrated a typical logistic model plot below, as you can you see the probability stays between 0 and 1 at all times.

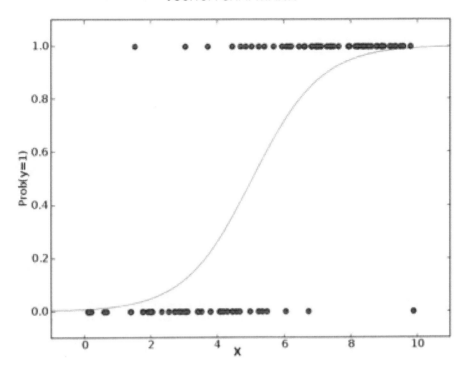

Performance of Logistic Regression Model

In order to evaluate the performance of a logistic regression model, there are a few metrics that you will need to consider. Regardless of the tool which you are working on, whether it is SAS, R or Python, you will need to look for the following:

1. AIC (Akaike Information Criteria) - This is the analogous metric of adjusted R2 in logistic regression and is known as AIC. The AIC is the measure of fit which penalises based on the number of model coefficients.

2. Null Deviance and Residual Deviance - Indicating the response predicted by the model, the Null Deviance uses nothing but an intercept. In this case the lower the value, the better the model. The residual deviance indicates a predicted

response by a model when adding independent variables meaning that the model is better when values are lower.

3. Confusion Matrix: Indicated below, the Confusion matrix is a tabular representation for actual vs predicted values. This will assist us in finding the accuracy of the model as well as avoid overfitting.

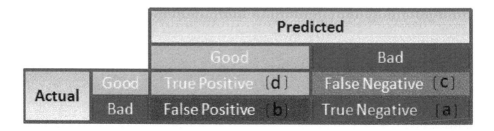

		Predicted	
		Good	Bad
Actual	Good	True Positive [d]	False Negative [c]
	Bad	False Positive [b]	True Negative [a]

You are able to calculate the accuracy of a particular model using the following equation

$$\frac{\text{True Positive} + \text{True Negatives}}{\text{True Positive} + \text{True Negatives} + \text{False Positives} + \text{False Negatives}}$$

When using a confusion matrix, you are able to derive Specificity and Sensitivity using the following equations:

$$\left. \begin{array}{l} \text{True Negative Rate (TNR), specificity } = \dfrac{A}{A+B} \\[3mm] \text{False Positve Rate (FPR), } 1 - \text{specificity } = \dfrac{B}{A+B} \end{array} \right\} \text{sum to 1}$$

$$\left. \begin{array}{l} \text{True Positive Rate (TPR), sensitivity } = \dfrac{D}{C+D} \\[3mm] \text{False Negative Rate (FNR) } = \dfrac{C}{C+D} \end{array} \right\} \text{sum to 1}$$

Both Specificity and Sensitivity have a significant role when deriving ROC curve.

4. ROC Curve: ROC stands for Receiver Operating Characteristic. What this does is summarises the performance of the model by weighing up the trade-offs between both the true positive rate (sensitivity) and the false positive rate (1-specificity). While plotting the ROC, it is best to assume that $p > 0.5$ for a greater focus on the success rate. What ROC does is summarizes the predictive power for all possible values of $p > 0.5$. The index of accuracy (A) or the concordance index is an ideal performance metric for ROC curve and can be found in the area of under the curve. The higher this area, the better we are able to predict using the model. You can see this in the sample ROC curve below. You will notice that the ROC of a perfective predictive model has TP equals 1 and FP equals 0. From this, we can see that the curve will touch the top left corner of the growth.

25

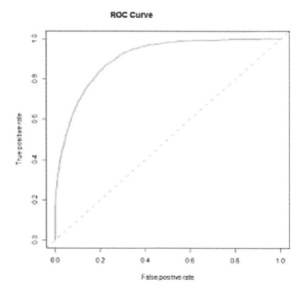

You may also like to consider the likelihood function for model performance. We call it the likelihood function due to the fact that it selects the coefficient values which maximise the likelihood of whether we can explain the data we are observing. As the value approaches one, we have an indication of how good of a fit it and the opposite is true of the data if the value approaches zero.

Chapter 6 – Unsupervised Learning: Support Vector Machines

Introduction

Machine learning algorithms may seem complicated at first however this isn't to say that cannot be mastered. Many newcomers to machine learning get started with simple algebraic regressions regress as they are simple to understand, apply and implement into your codes. However, there is much more you can do once you branch out and away from simple linear regressions.

One way to think of machine learning algorithms is as though they form parts of an armoury which is filled with all kinds of weapons such as axes, swords, blades, bows now if you were to think of regression as simply a sword that can slide and dice through the data, what about the cases where you will need to deal with more complex and precise data. In this case 'Support Vector Machine' are more like a sharp knife. They deal with much smaller datasets but they can be more powerful, stronger and effective when building models.

From our previous chapters, we covered some of the other models and now we are moving away from the basics and more towards some of the advanced concepts of machine learning algorithms, support vector machines.

What is Support Vector Machine?

"Support Vector Machine" (SBM) is used in classification and regression challenges as a supervised machine learning algorithm. A majority of the work is in classification problems however and using this algorithm, we plot each data item as a point in n-dimensional space. In this case, n is the number of features we have and the value of each feature will translate to the value of a particular coordinate.

The task is then to perform classification by determine the hyperplane that differentiates the two classes best.

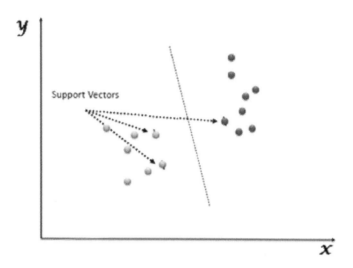

Support Vectors are basically just the coordinates of individual observation. Support Vector Machine is therefore a frontier which best separates each of the two classes, the hyperplane and line.

How Does It Work?

Now that we understand the process behind segregating each of the two classes using a hyper-plane, it is time to answer the question of how can we identify the right hyperplane? This might seem a little challenge but it is really quite simple.

We can do it through this process:

- Identify the right hyperplane. In this scenario we have three hyperplanes, A, B and C. We will need to identify the right hyperplane in order to classify the star and circle.

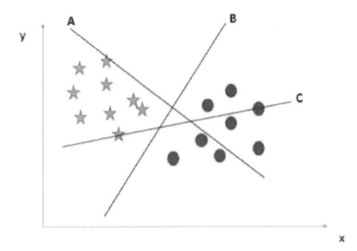

- There is a rule which you can remember in order to identify the right hyperplane and this is to select the hyperplane which segregates the two classes better. As we can see from this scenario, hyperplane B has done the best job in this case.

- In the next scenario, we have three hyperplanes, A, B and C however they are all segregating the classes in a similar way. How is it possible to identify the right hyperplane when neither stands out?

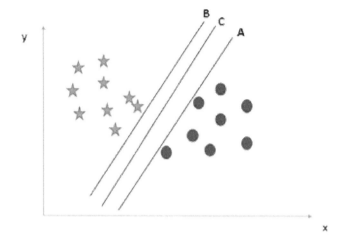

- In order to do this, we will use Margin. This is the distance between the nearest data point of either class and the hyperplane and will help us in deciding which is the right hyperplane. Let's have a look from the snapshot below.

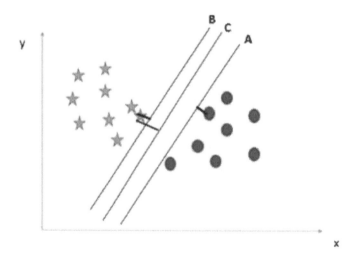

- As you can see in this image, the margin for hyperplane C is much higher compared to both A and B. In this case we would say that the right hyperplane would be C. The reason we use the hyperplane with the highest margin and this is due to the fact that a hyperplane with a low margin could lead to a high chance of misclassification.

- Have a look over the image below and using what you have learnt in the previous two scenarios to determine the right hyperplane. From just looking at the image, many of you might decide on hyperplane B having a much higher margin than 8. There is a little more to it than that. SVM selects the hyperplane which classifies the classes accurately before maximizing margin. In this case, Hyperplane B has an error it it's classification whereas A has classified much more accurately. From this, we determine that the right hyperplane is A.

- Now if we were to classify two classes due to the fact that we are unable to segregate any of the two classes using a straight line, what would happen then. As you can see from the image below, there is one star which is in the territory of the circle class as an outlier.

- This outlier for the star class is actually ignored by SVM as it must ignore any outliers and find the hyperplane that has maximum margin. Therefore, it is safe to say that SVM is robust outliers.

- Our next scenario is a little different. There is no way we can have a linear hyperplane segregating the two classes. In this case, how would SVM classify the two classes without a linear hyperplane.

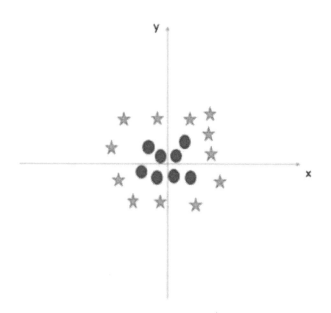

- This is actually quite a simple problem for SVM to solve. In order to do this, it will be introducing an additional feature into the problem. We do this by adding the feature $z=x^2+y^2$. From there we plot the data points on our chart at axis x and z.

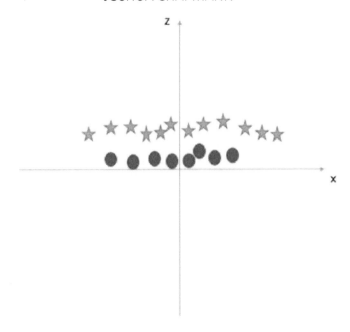

The points for us to consider when plotting the points:

a. All values for z will be positive. The reason for this is that z is constantly the squared sum of both x and y.

b.	In our original plot, we have red circles appearing close to the origin of x and y axes. This leads to our value of z to be lower and the star to be relatively away from the origin resulting in higher values of z.

In the case of SVM, it is much simpler to have a linear hyperplane segregating the two classes. We need to ask then however, in situations where we lack a hyperplane, does this feature need to be added manually in order to have one. The answer SVM brings us then is a technique known as the kernel trick. The kernel trick comprises of functions, known as kernels, that take low dimensional input space in order to transform it to a higher dimensional space such as converting a non-separable problem into a separable problem.

The kernel trick is best used in cases of non-linear separation problems as it does some extremely complicated data complications.

The advantage for us here is that the process is able to separate the data based on the labels or outputs that we have defined. As an example, we have the hyperplane below in the original space, you will notice that it looks like a circle.

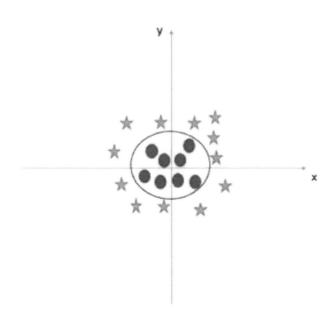

Pros and Cons Associated with SVM

Pros

- When there is a clear margin of separation, SVM can be highly effective.

- The effectiveness is increased in scenarios of high dimensional spaces.

- It is particularly effective in cases where there a greater number of dimensionals than the number of samples

- The technique utilises a subset of training points within the decision function, also known as support vectors. Meaning it is also memory efficient.

Cons

- In cases where there is a large dataset, SVM does not perform well. The reasoning for this is that the training time is much higher.

- In cases where there are many target classes overlapping and there is a lot of action, SVM does not perform too well.

- SVM fails to directly provide probability estimates. Instead these are calculated using fivefold cross validation which can be quite expensive as a result. This is a related SVC method of Python scikit-learn library.

Chapter 7 – Unsupervised Learning: Advanced Neural Networks

This next algorithm, like many of the others, may seem a little intimidating at first glance with the tagline for this algorithm being "It works in a way, similar to the human brain". Hearing this for the first time, you can be forgiven thinking this algorithm is impossible complex like the human brain but the Advanced Neural Networks algorithm is actually quite simple once you dive into it a little further.

Imagine for a moment that every human brain was as simple as the CPU of a desktop. What would be different. Your answer would probably be that most of us would behave exactly the same with every situation having an identical reaction, leading to the same decisions and ultimately the same outcomes. What is it that makes our brain so complex and individual to the point where this doesn't happen? There is one answer the author can give and that is that a machine uses simple algorithms.

These algorithms forming a simple machine, convert inputs to outputs. In this case every same input will always lead to the same output. In the case of our human brain, we have the unique characteristic of creating transient states through neurons in between our sensory organs and our braining, where our decisions come from. What happens then is a probabilistic interim state which can add a factor of randomness to scenario, this is referred to in our society as creativity.

In the case of an Artificial Neural Network (ANN) or any machine learning algorithm for that matter, we would need to have constructed a similar transient state in order to allow for the machine to learn in a more complex manner. Throughout this chapter, we want to draw comparisons of the framework of ANN algorithm with the functionality of the human brain.

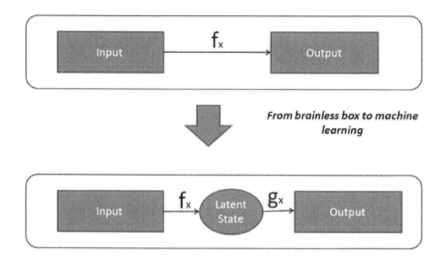

How does a simple predictive algorithm work?

What those simple predictive algorithms attempt to do is mimic the relationship between the input and output variables. The function derived in these scenarios is one which is direct linear or nonlinear between input and output variables. As an example, if we were to try to predict the amount of total work experience of person if the information available to us was their age, we would have this kind of relationship plotted.

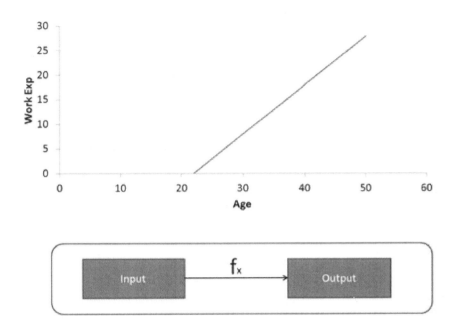

You are able to simply predict these kinds of relationships using nothing more than simple regression algorithms. However, this can become rather difficult when trying to make a prediction in cases of complex nonlinear relationships and significant covariate terms. Therefore, in order to make a prediction, there are two options available to us. We can either predict a complex nonlinear function or we can choose to divide this problem up into multiple, smaller steps and solve for each step. We can do this through using an artificial neural network. (ANN).

How does ANN Work?

Going back to our introduction at the start of this chapter, the working of ANN really does have some strong similarities from the neural network of our very own human brain. ANN operates on what has come to be referred to as Hidden State. These hidden states are where we get our comparison with the neurons in our brains. Each hidden state is a transient form of which it has a probabilistic behaviour similar to how our brain would operate. Acting as the bridge between the input and output is a grid of these hidden states as we can see in the diagram below.

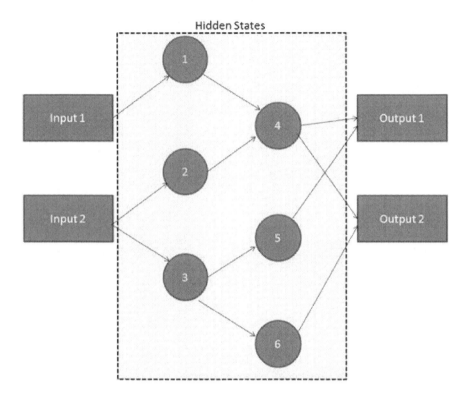

Let's analyse this diagram above carefully to fully understand what it means to us. Firstly, we have a vector comprised of three inputs and what we will need to do from there is find the probability that the out event will fall into either class 1 or class 2. In order to make this prediction, we will need to predict a series of hidden classes found between the bridge. The vector of the three inputs will use a probabilistic approach to predict a combination based on the activation of hidden nodes from 1 - 4. From there, the hidden nodes of 5-8 will then be used to predict the hidden nodes of 9-12. Once that is done, we will be able to finally predict our outcome. The intermediate latent states will allow our algorithm to determine the results of every prediction.

Take a Look on Different Scale

With an understanding of how neural networks now in your grasp, there are still some questions that you might have regarding the application and usage. Neural networks are currently leading the way in machine learning and researching today with many of the largest data giants such as Google, Facebook, Baidu and others investing heavily in researching how these algorithms work and how to best utilise them. An example of how this can be done would be solving voice recognition problems. Since all the past dialogues spoken by you would have become an input for a neural network, the words in the dictionary become the output. With a large enough number of outputs across several people, we are able to effectively create a neural net which is able to predict the correct words as the output. This kind of work has become the basis for research labs across the world. The subject has also been come to be known as deep learning.

Endnotes

From this article, you are able to gain some insight into a concept that can seem very murky and complex known as Artificial Neural Network. The name alone can create many misconceptions for people but having read through this chapter, you should now be a little more curious and hopefully desire to learn far more about this algorithm. We merely covered the tip of the iceberg through this chapter and there are many resources available to you if you feel like to go further in detail towards learning more about the algorithm.

You might also like to learn more about how the algorithm is able to learn as well as the linkages between the input and hidden state or even amongst the hidden state and also between the hidden state and output. There is also much to learn about the feedback learning which is also known as backward learning mechanism which assists in optimizing the linkage grid between all the hidden states. These concepts might seem very foreign to you at this stage but it is actually quite simple once you begin learning and as this is one of the most rapidly growing subjects in machine learning, you have many benefits to take advantage of from learning more about neural networks as well as deep learning.

Chapter 8 – Unsupervised Learning: Decision Trees

If you have ever worked in a business analyst role within an organization, you will notice that there is a preference or bias for a particular classification technique, this being the decision tree. The decision tree is preferred mainly due to its inherent simplicity and numerous advantages. We will cover the decision tree a little further throughout this chapter. Decision trees are found to be one of the most common classification tools in business analyst positions. Not only can a decision tree help us with prediction and classification but it is also an incredibly effective tool for underlying the behaviour of various variables. We will be covering the algorithm in detail throughout this chapter.

What is a Decision Tree?
A decision is a type of supervised learning algorithm. It needs to have a predefined target variable and the tool is mostly used within classification problems. The decision tree works for both categorical as well as continuous input and output variables. The technique works like this. We split up the population or sample into or more homogeneous sets or sub populations and we base this on the most significant splitter or differential in input variables.

As example, let's say that we are working with a sample of 30 students and each has three variables these being gender, whether they are a boy or a girl, Class, this case we will say IX and X and height, 5 to 6 feet. Now we have 15 out of these 30 kids play cricket during their leisure time. In order to create a model to predict who will be playing cricket during the leisure period, we will need to overcome a problem. The problem is that we will need to segregate the students who play cricket during their leisure time based on highly significant input variable among all three of the variables. A decision tree in this situation, will be able to help us solve this problem. What the decision tree does is segregate the students based on all values of three variables as well as identify the variable which creates the best homogeneous sets of students despite the fact they are all heterogeneous to each other.

We are able to see this in the snap show below. You will notice that the variable generable is the variable to best identify the ideal homogeneous sets compared to the other variables of class and height.

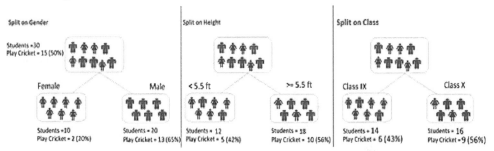

As we can see from these figures, the decision tree is able to identify the variable that holds the most significance as well as the one which is able to create the most homogeneous population set. This then leaves an unanswered question which is, how does the decision tree identify the variable and the split. The answer to this question is that the decision makes use of various algorithms, while we won't be covering these algorithms extensively in this chapter, I urge you to do some further research if you are curious in order to learn how decision trees work.

Types of Decision Tree

There are two types of decision tree and the one you use will depend largely on the type of target variable you are using. The two types of decision trees are:

1. Binary Variable Decision Tree: As the name suggest, the Binary Variable Decision Tree is used in cases where the Decision Tree has a binary target variable. For example, going back to our scenario with our students, the target variable in question was whether the student will play cricket or not in which case we have two answers, YES or NO.

2. Continuous Variable Decision Tree: In other scenarios where the Decision Tree has continuous target variable then you will use a Continuous Variable Decision Tree. An example of how this would be used would be if we were faced with a problem to predict whether one of our customers was going to pay his renewal premium through our insurance company or not. There are significant variables we are presented with here such as the customer's income, however our insurance company does not have the details of income for all our customers. In this case it is a very important variable so we will need to build a decision tree in order to predict the customer's income based on his occupation, product he is using as well as any other variables that we are available to us. This is what we mean when we say we are predicting variables for continuous variable.

Terminology Related to Decision Trees

There are few basic terminologies you will need to learn in order to have a greater understanding of decision trees. These are:

ROOT Node: This represents the entire population or sample and is becomes further divided into the varying sets by the decision trees to form more homogeneous sets.

SPLITTING: This is the process in which a node is then divided into two or more sub-nodes

Decision Node: After a split has occurred in a sub-node to further divide it into sub nodes, this is known as the decision node.

Leaf/Terminal Node: In the case where a node is not split, it becomes known as the Leaf or Terminal Node.

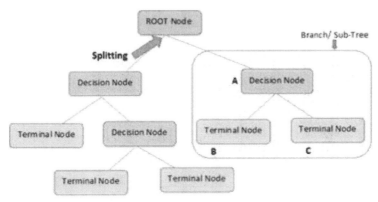

Note:- A is parent node of B and C.

Pruning: The act of removing sub-nodes from a decision node is known as pruning. This can have thought of to be the opposite process of splitting.

Branch/Sub-Tree: When we divide our tree into entire sections, these sections are then called branches or subtrees.

Parent and Child Node: When we have a node, which is then divided into a sub node we then have the parent node of sub nodes whereas our sub nodes become known as the child of a parent node.

These terms will help you concerning the concept of decision trees as they are quite commonly used. It is important to understand that every algorithm has its advantages and disadvantages and situations where it is better suited for solving the problem and the decision tree is no exception. Below are the pros and cons of decision trees as an algorithm.

Advantages

1. Simple to Understand: Decision tree outputs are one of the easiest algorithms to understand. Even those who are from a background outside analytics, they can still find Decision trees simple to pick up and use. In order to get started using Decision Trees, you don't need to have any prior statistical knowledge in order to analyse and interpret them. With the graphical representation, many intuitive users are able to easily understand and relate their hypothesis.

2. Useful in Data Exploration: When identifying more significant variables and determine the relationship between two or more variables, Decision Trees are one of the fastest ways we are able to do this. Decision Trees can also assist us in creating new variables and features in order to have better power to predict the target variable. For example, if we are working with a problem that gives information filled with hundreds of variables, we are able to use a decision tree in order to identify the most significant variable.

3. Less Data Cleaning Required: Decision trees also require less data cleaning in comparison to some of the other modelling techniques we have explored as it is not heavily influenced by outliers or any values that may be missing.

4. Data Type is not a constraint: Decision trees can handle both numerical and categorical variables

5. Non-Parametric Method: The decision tree is considered to be a non-parametric method. The reasoning for this is that decision trees have no assumptions about both the classifier structure as well as the space distribution.

Disadvantages

1. Overfit: Decision Trees have difficulty in Overfitting and it is simply not practical. WE are able to solve this issue using random forests however this requires some further research and is a little complex for this book.

2. Not Fit for continuous variables: When we are working with continuous numerical variables, our decision tree loses some of the information in the process of categorizing the variables into different categories.

End Notes

Throughout this chapter, we explored one of the more well known and most commonly used techniques in predictive modelling and exploratory analysis. As you can see, this method can be incredibly effective particularly in rapid prototyping of models. Decision trees are great for beginners, especially those looking at learning algorithms yet don't want to jump into the deep end and can be a great way to explore further topics within machine learning.

22446723R00031

Printed in Great Britain
by Amazon

Table of contents

Chapter 1 – Introduction to Machine Learning

To make any computer program work, you must build a set of instructions that tell it exactly what to do; these are called the *code*. Here you must carefully define every input, every calculation and every output, because computers are unable to figure things out on their own. You can write thousands of lines of code perfectly, but if there is a single comma missing your program will crash.

In these occasions a human person must read the code, assess what the problems are, identify the missing comma and insert it – this process is called *debugging*. Why can't computers do this process by themselves? Simple, because no one programmed them to. The ability to analyse new scenarios and adapt to changing situations is called *learning*. It is only found in intelligent creatures, not computers.

Machine Learning is a branch of computer science that wants to change this fact. They want to stop giving computers detailed instructions and instead provide them with a high-level set of commands which they can apply to many different scenarios – these are called **algorithms**. In practice, they want to give computers the ability to **learn** and to **adapt**.

We can use these algorithms to obtain insights, recognize patterns and make predictions from data, images, sounds or videos we have never seen before (or even knew existed). Unfortunately, the true power and applications of today's Machine Learning Algorithms is misunderstood by most people.

Through this book I want fix this confusion, I want to shed light on the most relevant Machine Learning Algorithms used in the industry. I will start by discussing popular algorithms for supervised learning

1

(KNN, Naïve Bayes and Regressions) and then move onto the more complex field of unsupervised learning (Support Vector Machines, Neural Networks and Decision Trees). I will discuss exactly how each algorithm works, why it works and when you should use it.